NATIONAL GEOGRAPHIC

Ladders

TRICKS, TRAPS, AND TOOLS

Tricks

by Judy Elgin Jensen

Imagine walking deep in a rain forest in Queensland, Australia. Dead leaves litter the ground—shades of brown in all different shapes. But look closely. Is everything you see *really* a leaf? In nature, what you see is not always what you get!

Among the leaves are several leaf-tailed geckos. Their colors and shapes are like those of the leaves. Even their tails look like leaves. The geckos are very still, waiting to ambush unwary insects. Their stillness also makes them nearly invisible to **predators** that would like to eat them, such as owls and snakes.

How many leaf-tailed geckos can you find? The background has been lightened to make it easier to see one of them.

While they are on the forest floor, the colors of the leaf-tailed geckos are similar. But soon they will scamper away from one another, climbing up different trees. There each gecko's coloring will change to other shades of brown or green. Even though they are all the same kind of gecko, they will then look very different from one another.

Have you found the leaf-tailed geckos yet? If you look closely, you can find three of them.

Hiding in Plain Sight

Like leaf-tailed geckos, many animals rely on **camouflage** for protection. The coverings of these animals—their skins, shells, fur, or feathers—look like their surroundings. It is difficult for a predator to see an animal that is the same color and shape as the things around it. Some predators use camouflage to hide from their prey. If their prey wander too close, they quickly catch and eat them.

Jagged Ambush Bug
Where it lives: Eastern North America
What's special: This bug sits and waits on flowers, especially goldenrod. When an insect lands nearby, the bug grabs it and injects poison into its body. The bug often ambushes insects much larger than itself. To find the bug, look for its green and brown body.

Malaysian Orchid Mantis
Where it lives: Rain forests in Malaysia
What's special: This mantis looks like an orchid blossom. Some of these insects are pink or purple, just as some orchids are pink or purple. To find the mantis, look for its dark eyes.

Soft Coral Crab
Where it lives: Coral reefs near Indonesia
What's special: This crab lives among coral polyps. To find the crab, look for its spiny legs with pink stripes.

Looking Dangerous

Not all animals survive by camouflage. Animals that are dangerous often have bright colors or patterns that make them stand out from their surroundings. Predators avoid animals with warning coloration. Some harmless animals have colors and patterns that mimic, or copy, those of dangerous animals. These copycats stay safe through the protective trick of **mimicry**.

Model: Plain Tiger Butterfly

As a caterpillar, the plain tiger eats leaves that make the adult taste bad to most predators. The adults can also release a bad-smelling fluid.

Mimic: Danaid Eggfly Butterfly

The female danaid eggfly looks like the plain tiger, but does not taste bad. Predators mistake this butterfly for the bad-tasting plain tiger butterfly.

Model: Yellowjacket

Like other wasps, the yellowjacket has a stinger that can hurt predators.

Mimic: Yellowjacket Hover Fly

This harmless hover fly looks like a yellowjacket. It sounds like one, too.

Model: **Coral Snake**

The venomous coral snake has bands of yellow, black, and red. Like a bright red fire truck speeding down the street, its bands are saying, "Watch out!" The scarlet king snake is not venomous, but has bands that look like those of the coral snake. Predators have learned that coral snakes can make them sick. So they avoid all banded snakes, including the scarlet king snake.

Mimic: **Scarlet King Snake**

Confusing Predators

Some animals behave in ways that confuse their predators or even scare them away. They may puff themselves up to look bigger or reveal markings that make them look like another, larger animal.

Io Moth
When startled, the io moth (above) spreads its wings. Its hindwings have large black eyespots. To many predators, these eyespots look like the eyes of an owl (to the right).

Puss Moth Caterpillar
The caterpillar of the Euoprean puss moth is about the same color as a leaf. But when it is startled, it pulls its head into its body and rears up. A predator sees a big red "mouth." If that doesn't scare the predator away, the caterpillar's two tails squirt acid at it!

Some animals have camouflage that keeps them hidden from predators. Other animals have bright colors and patterns that keep predators away. But whatever the strategy, the goal is always the same—survival.

Check In How does mimicry protect an animal from predators?

GENRE Reference Article

Read to find out how some plants get nutrients in an unusual way.

Traps

by Judy Elgin Jensen

A small frog hops among the sticky sundew plants in its boggy home, never suspecting that the plants are dangerous. Oops! Too close. Now the frog is stuck to a plant. Eventually the sticky liquid of the sundew will smother the frog. Then the frog will be turned into a nutritious soup for the sundew plant.

Plants use sunlight to make their own food. But plants also need **nutrients,** such as nitrogen, which they take in from the soil. The soil of many swamps and bogs does not contain enough nutrients. **Carnivorous plants** that live in these places get their nitrogen by trapping small animals.

This frog is caught in a sticky sundew plant.

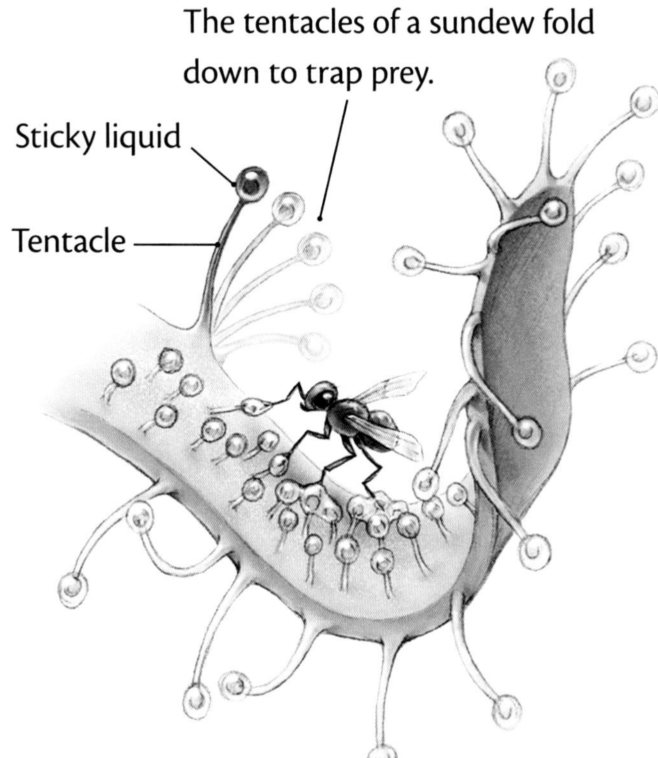

Sundews

There are about 150 different kinds of sundew plants. Some are as tiny as a dime. Others are taller than you are! The leaves of some sundews grow in a circle, while others stretch upward in stalks. Thick hairs, or tentacles, cover the leaves. Each tentacle has a drop of sticky liquid that holds the animal tight and digests it, too.

Bladderworts

A netlike plant covered with tiny "bubbles" floats in a pond. As a small animal swims by, it touches trigger hairs on one of the bubbles. WHOOSH! The water around the animal is sucked into the bubble, like water going down a toilet. The water carries the animal with it. That's no bubble. It's a trap!

The trap slams shut. Digestive juices fill the trap. Goodbye tiny animal.

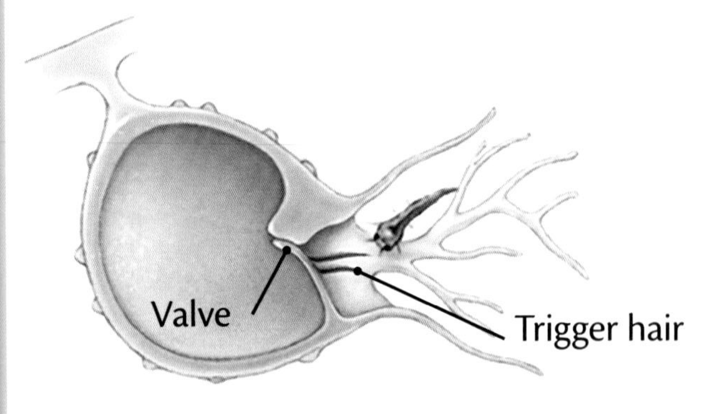

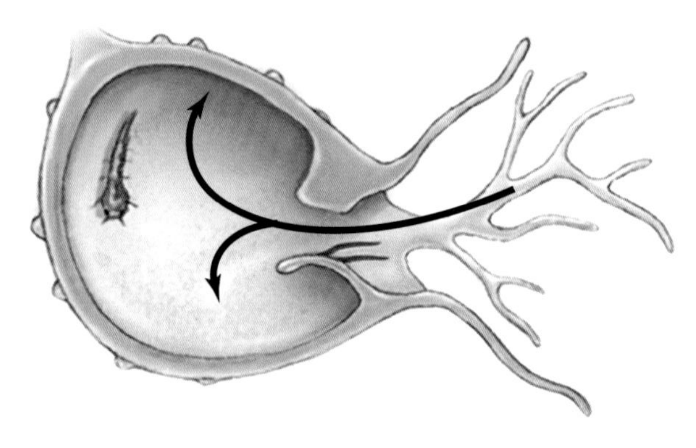

The bladderwort pumps water out of its trap. If an animal touches a trigger hair, the trap opens. Water rushes in. The animal goes in, too.

Bladderworts don't have roots to get nutrients from soil. Instead they get their nutrients from tiny animals.

Venus Flytraps

Venus flytraps capture flies and other animals that creep across their odd-looking leaves. On the inside of each leaf are trigger hairs. If an animal touches them, the trigger hairs send a signal to the leaf. The leaf snaps shut, trapping the animal. After about seven days, the trapped animal is completely digested. Then the leaf opens again, ready for its next victim.

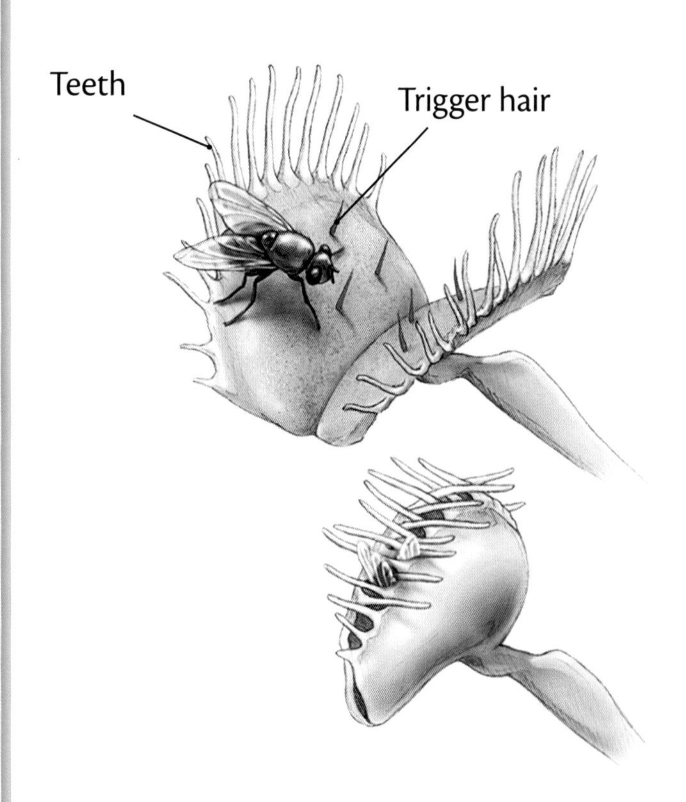

When an insect walks across a Venus flytrap leaf, it touches the trigger hairs. That causes the two halves of the leaf to snap shut.

Pitcher Plants

The sweet smell of nectar wafts through the air from a strange, pitcher-shaped leaf. An ant leans in to sip the nectar, but soon finds itself sliding down the slippery inside of the pitcher. At the bottom is a pool of water filled with digestive fluid. There the ant drowns. As the ant's body breaks down, it releases nutrients that the plant needs.

As the ant tries to crawl out of the pitcher, it runs into hairs that point downward. The hairs act as spears and keep the ant in the pool.

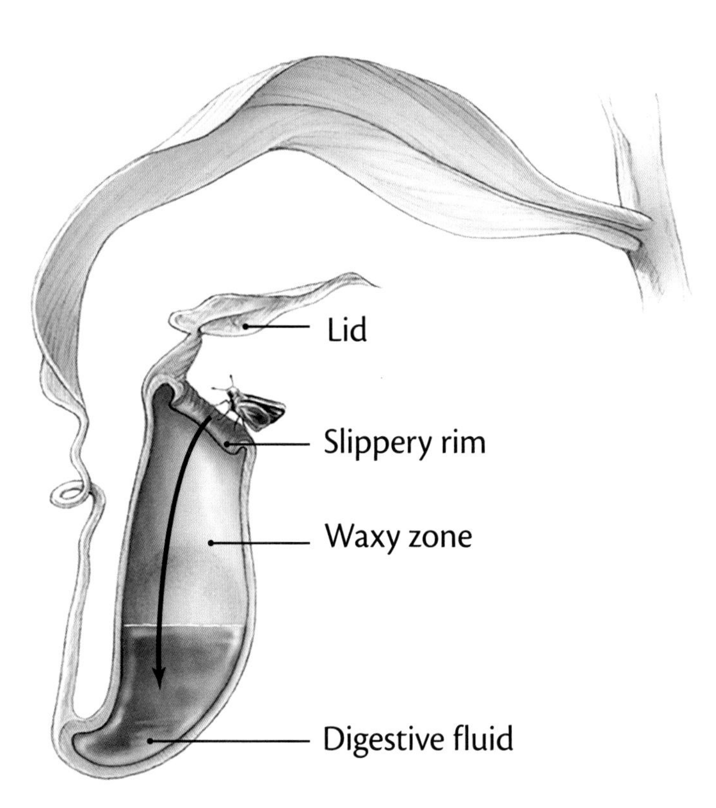

Carnivorous plants thrive where most other plants cannot get enough nutrients to survive. Their specialized structures trap and digest small animals, giving the plants the nutrients they need to live.

Yum. Bug juice.

Check In How does the leaf of a pitcher plant trap an ant?

GENRE Science Article **Read to find out** some ways that animals use tools.

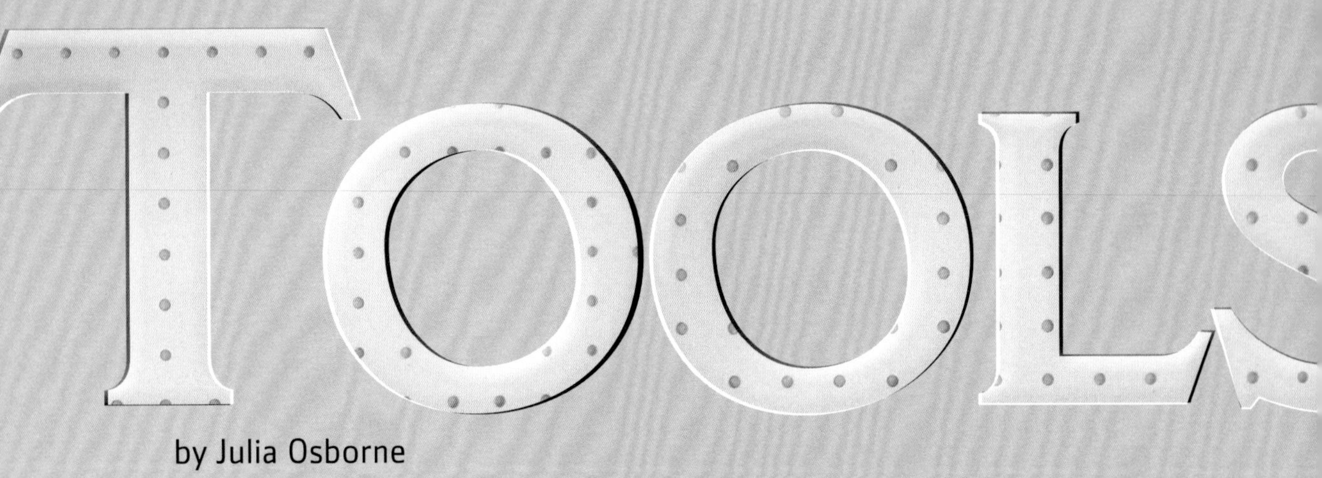

TOOLS

by Julia Osborne

What kinds of tools do you use? You probably use many, from simple ones, such as spoons, to complex ones, such as computers. A tool is an object used to carry out a task. Humans use a great variety of tools.

Scientists once thought only humans had enough brainpower to use tools. But that idea is changing.

For example, look at the Egyptian vulture. This bird likes to eat eggs. If an egg is small enough, the vulture will pick it up and drop it. When the egg breaks, the vulture slurps up the insides. Ostrich eggs are too big for the vulture to pick up. Instead it breaks them by throwing stones at them. When a vulture uses a stone to break an egg, the stone is a tool.

Read on to learn how several kinds of animals use objects to perform a task or change their environment.

An Eyptian vulture drops a stone on an egg.

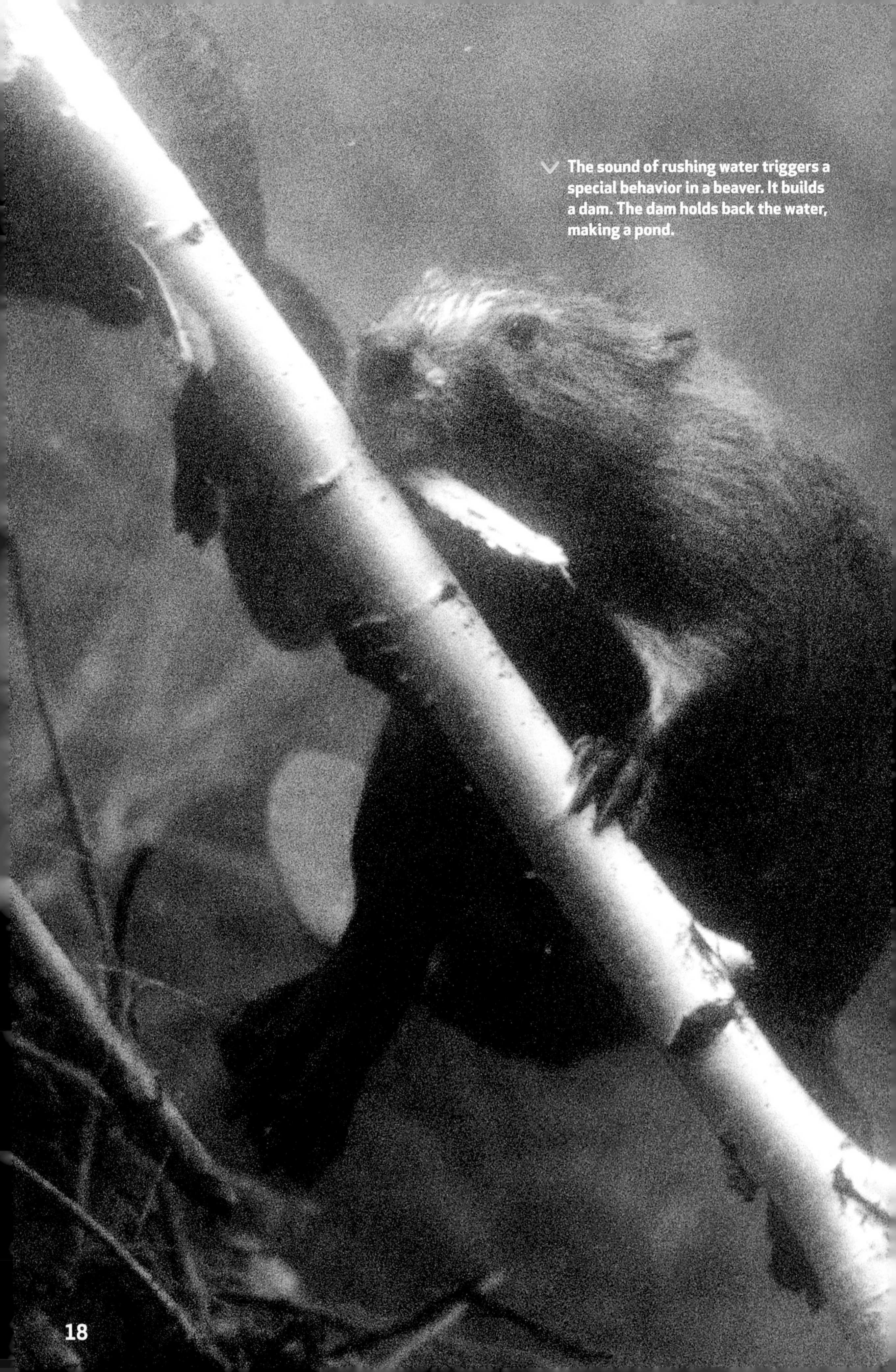

The sound of rushing water triggers a special behavior in a beaver. It builds a dam. The dam holds back the water, making a pond.

Using Instinct

All animals carry out complex actions called **behaviors.** Some animals act by **instinct,** without learning how or being taught by other animals. Animals inherit instincts from their parents.

To see how instinct can guide behavior, look at the work of beavers. Beavers cut down trees. Then they use the branches, sticks, and mud to build lodges in ponds. Inside their lodges, the beavers are safe from predators.

What if there is no pond? Beavers use sticks and mud to build a dam across a stream. The dam holds the running water and creates a quiet pond. The sticks and mud are tools that change the beavers' environment.

Do young beavers learn how to build dams from adult beavers? To find the answer, scientists watched young beavers that had been raised apart from adult beavers. On their first try, the youngsters built a dam that was just like the dams built by beavers in the wild. The answer to the question was clear: Building dams is an instinct.

> This capuchin monkey is cracking a nut.

Learning to Use Tools

As a young child, you learned how to use a spoon to eat. Many young animals learn how to use tools, too. Wild capuchin (KAP-yu-chin) monkeys learn how to use stones to crack open nuts and get to the tasty meat inside. Cracking nuts is a learned behavior.

Almost from birth, young capuchins are very interested in the nut-cracking behavior of other monkeys. But it takes a few years of observing and then trying on their own before they get the hang of it. Infant monkeys beat stones and other objects against things. One-year-olds start trying to crack nuts. By about age three, they figure out where to place the nut and how to accurately drop the stone. They have become expert nut crackers!

Sometimes adult monkeys who don't know how to crack nuts come into the group. These adults learn the skill by watching other monkeys and by practicing on their own.

A sea otter uses a rock to open a clam.

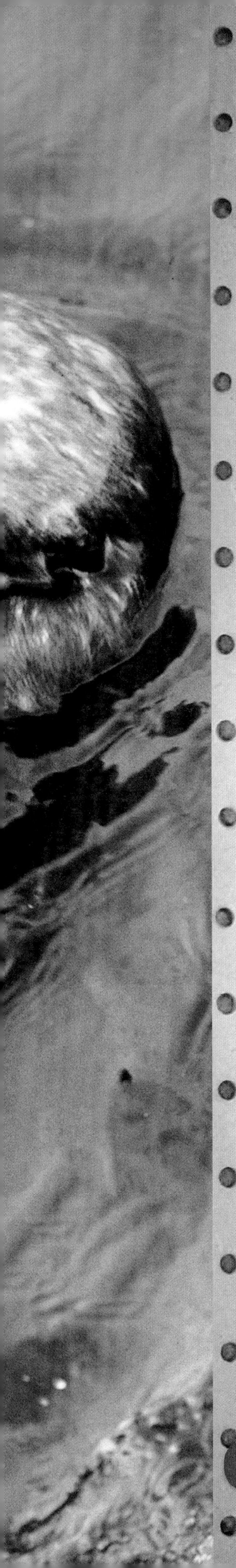

Water Crackers

Furry sea otters spend almost all of their time in the ocean—hunting, sleeping, and eating. They often float on their backs, holding themselves in place with long strands of seaweed.

Some sea otters have learned how to use rocks as tools. When one of these otters gets hungry, it dives to the ocean floor to get a clam. It also picks up a rock.

Floating on its back, the otter places the rock on its belly. Then it smashes the clam against the rock. Soon the clamshell cracks open. U-m-m-m-m. A tasty morsel.

No wonder scientists are changing their ideas about how animals use tools. They have discovered that some animals know how to use tools by instinct. Others learn to use tools by watching adults or by trying the skill on their own.

Check In What is the difference between an instinct and a learned behavior?

Discuss

1. The three pieces in this book are "Tricks," "Traps," and "Tools." Describe some of the ways these three pieces are connected.
2. Think about the animals in "Tricks." What are some ways that their shapes and colors protect them from predators?
3. Compare the actions of the sundew plant in "Traps" with the behavior of the ambush bug in "Tricks." How are they alike and different?
4. Describe how capuchin monkeys learn to open nuts.
5. What else would you like to know about the plants and animals in this book? How could you find out more?